V. I. Arnold

Catastrophe Theory

With 65 Figures

Translated from the Russian
by R. K. Thomas

Springer-Verlag
Berlin Heidelberg New York Tokyo
1984

Vladimir Igorevich Arnold
Department of Mathematics
University of Moscow
Moscow 117234
USSR

ISBN 3-540-12859-X
Springer-Verlag Berlin Heidelberg New York
Tokyo
ISBN 0-387-12859-X
Springer-Verlag New York Heidelberg Berlin
Tokyo

Library of Congress Cataloging in Publication Data. Arnol'd, V. I.
(Vladimir Igorevich), 1937 –. Catastrophe theory. Translation of: Teo-
riíà katastrof. Includes bibliographical references. 1. Catastrophes
(Mathematics) I. Title. QA614.58.A7613, 1984. 514'.7. 83-20031
ISBN 0-387-12859-X (U.S.)

© Springer-Verlag Berlin Heidelberg 1984
Printed in Germany

Typesetting, printing and bookbinding: Zechnersche Buchdruckerei,
Speyer
2141/3140-543210

To the Memory of
M. A. Leontovich

Preface to the English Edition

Singularity theory is growing very fast and many new results have been discovered since the Russian edition appeared: for instance the relation of the icosahedron to the problem of by-passing a generic obstacle. The reader can find more details about this in the articles "Singularities of ray systems" and "Singularities in the calculus of variations" listed in the bibliography of the present edition.

Moscow, September 1983   V. I. Arnold

Preface to the Russian Edition

"Experts discuss forecasting disasters" said a New York Times report on catastrophe theory in November 1977.

The London Times declared Catastrophe Theory to be the "main intellectual movement of the century" while an article on catastrophe theory in Science was headed "The emperor has no clothes".

This booklet explains what catastrophe theory is about and why it arouses such controversy. It also contains non-controversial results from the mathematical theories of singularities and bifurcation.

The author has tried to explain the essence of the fundamental results and applications to readers having minimal mathematical background but the reader is assumed to have an inquiring mind.

Moscow 1981   V. I. Arnold

Contents

Chapter 1. Singularities, Bifurcations, and Catastrophe Theories

The first information on catastrophe theory appeared in the western press about ten years ago. In journals like "News Week" there were reports of a revolution in mathematics comparable with Newton's invention of the differential and integral calculus. It was claimed that the new science, catastrophe theory, is much more valuable to mankind than mathematical analysis. While Newtonian theory only considers smooth, continuous processes, catastrophe theory provides a universal method for the study of all jump transitions, discontinuities, and sudden qualitative changes. There appeared hundreds of scientific and popular science publications in which catastrophe theory was applied to such diverse fields as, for instance, the study of heart beat, geometrical and physical optics, embryology, linguistics, experimental psychology, economics, hydrodynamics, geology, and the theory of elementary particles. Among the published works on catastrophe theory are studies of stability of ships, models for the activity of the brain and mental disorders, for rioting prisoners and for investors on the stock exchange, studies of the effects of alcohol on drivers and of the censor's attitude towards pornographic literature.

In the early seventies catastrophe theory rapidly became a fashionable and widely publicized theory not unlike the all-embracing pseudo-scientific theories of the past century.

The mathematical articles of the founder of catastrophe theory, René Thom, were reprinted as a pocket book – something that had not happened in mathematics since the introduction of cybernetics from which catastrophe theory derived many of its advertising techniques.

Sine the eulogies of catastrophe theory, there have appeared more sober critical works. Some of these also appeared in publications with a wide readership, under eloquent titles like – 'The Emperor has no clothes'. Now we have articles devoted to criticism of catastrophe theory. (See for instance John Guckenheimer's survey article 'The Catastrophe Controversy' in 1978 and the parody on the criticism of the theory in 1979.)*

The origins of catastrophe theory lie in Whitney's theory of singularities of smooth mappings and Poincaré and Andronov's theory of bifurcations of dynamical systems.

Singularity theory is a far-reaching extension of the study of functions at maximum and minimum points. In Whitney's theory functions are replaced by mappings, i.e. collections of several functions of several variables.

The word *bifurcation* means *forked* and is used in a broad sense for designating all possible qualitative reorganizations or metamorphises of various objects resulting from changing the parameters on which they depend.

Catastrophies are violent sudden changes representing discontinuous responses of systems to smooth changes in the external conditions. In order to understand what catastrophe theory is about one must first become acquainted with the elements of Whitney's singularity theory.

* Guckeinheimer, J.: *The Catastrophe Controversy,* The Mathematical Intelligenser 1978, vol. 1, no. 1, pp. 15–21; Fussbudget H. T. and Znasber R. S., *Sagacity Theory: A Critique,* loc. cit. 1979, vol. 2, no. 1, pp. 56–59.

Chapter 2. Whitney's Singularity Theory

In 1955 the American mathematician Hassler Whitney published the article 'Mappings of the plane into the plane' laying the foundations for a new mathematical theory of singularities of smooth mappings.

A *mapping* of a surface onto a plane associates to each point of the surface a point of the plane. If the point on the surface is given by co-ordinates (x_1, x_2) and the point on the plane by (y_1, y_2), then the mapping is given by a pair of functions $y_1 = f_1(x_1, x_2)$ and $y_2 = f_2(x_1, x_2)$. The mapping is said to be *smooth* if these functions are smooth (i.e. differentiable a sufficient number of times, such as polynomials.)

Mappings of smooth surfaces onto the plane are all around us. Indeed the majority of objects surrounding us are bordered by smooth surfaces.

The visible contours of the object are the projections of the bounding surface onto the retina of the eye. By examining the objects surrounding us, for instance, people's faces, we can study the singularities of visible contours.

Witney observed that generically* only two singularities are encountered. All other singularities disintegrate under small movements of the body or of the angle of the projection while these two types are stable and persist after small deformations of the mapping.

An example of the first kind of singularity, called a Whitney *fold* is the singularity arising at equatorial points when a sphere is projected onto a plane (Fig. 1). In suitable co-ordinates, this mapping is given by the formulae $y_1 = x_1^2$, $y_2 = x_2$.

---

* "Generically": means for all cases bar some exceptional ones.

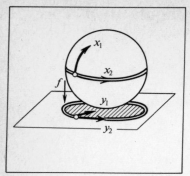

Fig. 1

The projections of surfaces of smooth bodies onto the retina points always have such singularities and there is nothing surprising in this. What is surprising is that besides this singularity, the fold, we are encountering just one other type of singularity but practically never notice it.

The second Whitney singularity, the *cusp,* arises when a surface like that in Fig. 2 is projected onto a plane. This surface is given by the equation $y_1 = x_1^3 + x_1 x_2$ with respect to spatial co-ordinates (x_1, x_2, y_1) and projects onto the horizontal plane (x_2, y_1). In local co-ordinates the mapping is given by $y_1 = x_1^3 + x_1 x_2$, $y_2 = x_2$.

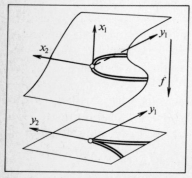

Fig. 2

Onto the horizontal plane one sees a *semi-cubic parabola* with a cusp (spike) at the origin. This curve divides the plane into two parts, the smaller and the larger. The points of the smaller part have three inverse images (three points of the surface project onto the point on the plane), points of the larger part only one and points on the curve two). On approaching the curve from the smaller side two of the inverse images (out of the three) merge together and disappear, (here the singularity is a fold), on approaching the cusp all three inverse images coalesce.

Whitney proved that the cusp is *stable*, i.e. every nearby mapping has a similar singularity at the appropriate point (that is a singularity such that, in suitable co-ordinates, a deformation/mapping in a neighbourhood of the appropriate point is described by the same formulae as those describing the original mapping).

Whitney also proved that *every singularity of a smooth mapping of a surface onto a plane, after an appropriate small perturbation, splits into folds and cusps.*

Thus the visible contours of generic smooth objects have cusps at points where the projections have cusp singularities and have no other singularities. Looking around we can find these cusps in the lines of every face or object. We consider for instance the surface of a smooth torus (an inflated tyre). The torus is usually drawn as in Fig. 3. If the torus were transparent we would see the visible contours indicated in Fig. 4. The corresponding mapping of the torus onto the plane has four cusp singularities, so the ends of the lines of the visible contours in Fig. 3 are cusps. At these points the visible contours have a semi-cubic singularity.

A transparent torus is rarely seen. We see other transparent objects – bottles, (preferably milk). In Fig. 5 two cusp points are visible. By moving the bottle a little we can deduce that the cusps are stable. So we have convincing experimental confirmation of Whitney's theorem.

After Whitney's basic work, singularity theory developed rapidly and is now one of the central areas of mathematics where the most abstract parts of mathematics (differential

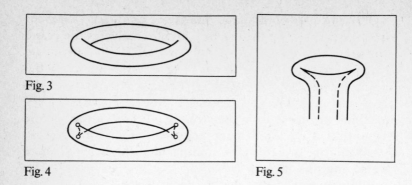

Fig. 3

Fig. 4

Fig. 5

and algebraic geometry and topology, theory of reflections, groups, commutative algebra, complex space theory, etc.) come together with the most applied (stability of dynamical systems, bifurcation of equilibrium states, geometrical and wave optics, etc.). René Thom suggested that the combination of singularity theory and its applications should be called catastrophe theory.

Chapter 3. Applications of Whitney's Theory

Since smooth mapping are found everywhere, their singularities must be everywhere also and since Whitney's theory gives significant information on singularities of generic mappings, we can try to use this information to study large numbers of different phenomena and processes in all areas of science. This sample idea is the whole essence of catastrophe theory.

When the mapping we are concerned with is known in detail we have a more or less direct application of the mathematical singularity theory to various natural phenomena. Such applications do indeed lead to useful results, such as in the theory of elasticity and in geometrical optics (the theory of singularity of caustics and wave fronts, about which we shall have much to say later).

In the majority of works on catastrophe theory, however, more controversial situations are considered where not only are the details of the mapping not known, but its very existance is problematical.

Applications of singularity theory in these situations are of a speculative nature, as an example of such an application we reproduce the one given more or less by the English mathematician Christopher Zeeman* of the application of Whitney's theory to the study of the activity of a creative personality.

---

* Zeeman, E. C.: Catastrophe theory: a reply to Thom, – in: Dynamical Systems, Warwick, 1974 Springer, Lecture Notes in Math., *468*, Berlin–Heidelberg–New York, 1975, p. 373.

We shall characterize a creative personality, (e.g. a scientist) by three parameters, called 'technique', 'enthusiasm', and 'achievement'. Clearly these parameters are inter-related. So we have a surface in three dimensional space with co-ordinates (T, E, A). We project this surface onto the plane (T, E) in the direction of A. For a generic surface the singularities are folds and cusps (Whitney's theorem). It is claimed that a cusp situated as indicated in Fig. 6 satisfactorily describes the observed phenomena.

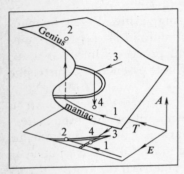

Fig. 6

We can see that under these assumptions the achievement of a scientist can be changed in relationship to its dependence on technique and enthusiasm. If enthusiasm is not great then achievement grows monotonically and fairly slowly with technique. If enthusiasm is sufficiently great then a different phenomena occurs. Now with increasing technique achievement can increase by a jump (such a jump occurs for instance at point 2 in Fig. 6 as enthusiasm and technique flow along curve 1). The domain of high achievement at which we have now arrived is indicated by the word 'genius'.

On the other hand a growth of enthusiasm not supported by a corresponding growth in technique leads to a catastrophe (at the point 4 of curve 3 in Fig. 6) where achievement falls by a jump and we drop to the domain denoted by the

8

word 'maniac'. We see that the jump from the state of genius to that of maniac and back goes along various lines so that for sufficiently great enthusiasm a genius and maniac can possess identical enthusiasm and technique, differing only in achievement (and previous history).

The deficiences of this model and many similar speculations in catastrophe theory are too obvious to discuss in detail. We remark only that articles on catastrophe theory are sharply distinguished by the catastrophic lowering of the level of the requirement of rigour and also of the requirement of the novelty of the published results. Although one can understand the catastrophe theorist's reaction against the traditional flow in mathematics of rigorous but dull epigonic works, nevertheless their lack of respect to their predecessors (to whom belong the majority of concrete results) can hardly be justified. The causes in both cases are more social than scientific*).

---

* "I think, my dearest, that all this decadence is nothing more than simply a way to approach tradesmen." V. M. Dorochevich, Rasskasy i Ocherki, Moscow 1966, p. 295.

Chapter 4. A Catastrophe Machine

In contrast to the example given above the application of singularity theory to the study of bifurcation of equilibrium states in the theory of elasticity is irreproachably founded.

In many elastic constructions under identical external loadings a number of different equilibrium states are possible. Consider, for instance, a horizontal rule whose ends are fixed by hinges and with a load at the centre. As well as the equilibrium state where the rule sags there is also the possible state where the rule is forced up into an arc like a bridge.

With increase of load, at some point a catastrophe or 'buckling' occurs. The rule suddenly jumps from one state to the other. Singularity theory has been applied to the study of these bucklings and its predictions have been well confirmed by experiment.

For visual illustration of this kind of application a number of devices have been invented. One of the simplest, Zeeman's Catastrophe Machine is illustrated in Fig. 7.

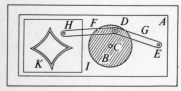

Fig. 7

The Catastrophe Machine can be easily made. One needs a board (A) (see Fig. 7) and a cardboard disk (B) secured to the board at its centre by a pin (C) so that it can freely rotate. Another pin (D) ist stuck into the disk at the edge and a third (E) into the board. To complete the assembly of the machine one needs two easily stretched rubber bands (F, G), a pencil (H) and a piece of paper (I).

After the pin on the edge of the disk has been linked to the fixed pin and the pencil by rubber bands, we place the point of the pencil somewhere on the sheet of paper so that the rubber bands are stretched. The disk settles itself in a certain position. Now as we move the pencil the disk rotates. It turns out that in certain positions of the pencil a small movement gives rise to a 'catastrophe' i.e. the disk suddenly jumps to a new position. If we mark all these 'catastrophe positions' on the sheet of paper, we get the 'catastrophe curve' (K).

It turns out that the catastrophe curve obtained has four cusps. On crossing the catastrophe curve a jump may or may not take place depending on how the pencil goes round the cusps.

By experimenting with this machine and trying to deduce the rule determining whether a jump takes place, the reader is easily convinced of the necessity of a mathematical theory of the phenomenum and more able to appreciate the value of singularity theory in this explanation.

The states of the catastrophe machine are described by three numbers. The position of the point of the pencil is given by two co-ordinates (called the *control parameters*). The position of the disk is determined by another parameter (the angle of rotation) called the *internal parameter* of the system. If all three parameters are given then the degrees of stretch of the bands and consequently the potential energy of the whole system are determined. The disk rotates until this energy is minimal (at least locally). For a fixed position of the pencil the potential energy is a function on the circle. This function can have one or several minima depending on the control parameters (Fig. 8, a). If on varying the control parameters the position of the minimum changes smoothly then there is no jump. A jump occurs for the values of the control parameters at which a local minimum disappears by combining with a

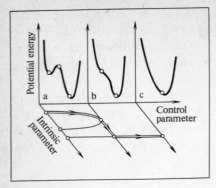
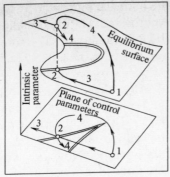

Fig. 8   Fig. 9

local maximum (Fig. 8, b); after the jump the disk arrives at a position determined by another local minimum (Fig. 8, c).

We examine the three-dimensional *state-space* of the machine. The states at which the disk is in equilibrium form a smooth surface in this space. We project this surface onto the plane of the control parameters (Fig. 9). This projection has folds and cusps. The points of the folds project onto the catastrophe curve. Fig. 9 shows clearly why a transition of the control parameters across the catastrophe curve sometimes causes a jump and sometimes does not. (It depends on which part of the surface the disk position is related to.) Using the diagram it is possible to get from one place on the surface to another without jumps occurring.

The situation for the majority of applications of catastrophe theory is similar to that of the examples described. We assume that the process under consideration is described by a number of control and internal parameters. The equilibrium states form a surface of some dimension in this space. The projection of the equilibrium surface onto the plane of control parameters can have a singularity. We assume that it is a generic singularity. Then singularity theory predicts the geometry of a catastrophe i.e. transition from one equilibrium state to another on changing the control parameters. In the majority of serious applications, the singularity is a

12

Whitney cusp and the result was known before the advent of catastrophe theory.

Applications of the type described are as well founded as are their original premises. For instance in the theory of buckling of elastic constructions and in the theory of capsizing of ships the predictions of the theory are completely confirmed by experiment. On the other hand, in biology, psychology, and the social sciences, for instances in application to the theory of the behaviour of market games or of nervous illness, as are the original premises, likewise the outcome is of rather heuristic value only.

Chapter 5. Bifurcations of Equilibrium States

An *evolutionary process* is described mathematically by a vector field in phase space. A point of phase space is called a *state* of the system. The vector at this point indicates the speed of change of the state.

At certain points the vector may be zero. Such points are called *equilibrium states* (states that do not change with time). Fig. 10 shows the phase space of the system describing hunters and hunted (say pike and carp). The phase space is the positive quadrant of the plane. The horizontal axis indicates the number of carp and the vertical axis the number of pike. P is the equilibrium state. The point A corresponds to the equilibrium number of carp with a less than equilibrium number of pike. Clearly oscillation will set in. The equilibrium state in Fig. 10 is *unstable*. The oscillation is represented by a closed curve. This curve is called a *limit cycle*.

Curves in phase space representing sequential states of a process are called *phase curves*. In a neighbourhood of a point that is not an equilibrium state, the partition of phase

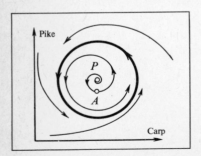

Fig. 10

space into phase curves is just like a partition into parallel lines. The family of phase curves can be transformed into a family of parallel lines by a smooth change of co-ordinates. In a neighbourhood of an equilibrium point the picture is more complex. It was shown by Poincaré that the behaviour of phase curves in a neighbourhood of an equilibrium point on the phase plane of a generic system is as in Fig. 11. All the more complex patterns disintegrate into combinations of these ones after a small generic perturbation.

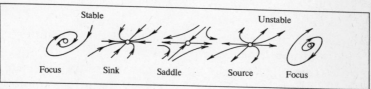

Fig. 11

Systems describing real evolutionary processes are as a rule generic. In fact such systems always depend on parameters that are never known exactly. A small generic change of parameters transforms a system that is not generic into one that is.

So, in general, situations more complex than those described are not encountered in nature and can be neglected in a first examination. This point of view displaces a large part of the theory of differential equations and of mathematical analysis in general where the main attention is traditionally paid to difficult non-generic special case studies of little real value.

However the situation is quite different if one is interested not in an individual system but in systems depending on one or more parameters. Then we consider the *space of all systems* (Fig. 12) divided into domains of generic systems. The dividing surfaces represent degenerate systems; under a small change of parameter values a degenerate system becomes non-degenerate. A one parameter family of systems is indicated by a curve in Fig. 12. This curve can intersect trans-

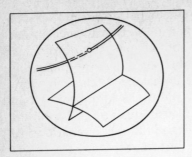

Fig. 12

versally (at a non-zero angle) the boundary dividing different domains of non-degenerate systems.

Thus although for each individual value of the parameter a system can be transformed by a small perturbation into a non-degenerate one, this cannot be done simultaneously for all values: every curve close to the one considered intersects the dividing boundary at a close value of the parameter (any degeneracy is removeable by a small perturbation at the given value of the parameter, but appears again at some nearby value).

And so *the degeneracies are not removeable when a whole family rather than an individual system is considered.* If the family is a one parameter family then the unremovable degeneracies are only the simplest being represented by boundaries of co-dimension one (i.e. given by one equation) in the space of all systems. One can save one parameter families from more complex degenerate systems forming a set of co-dimension two in the space of all systems by small perturbations.

If we are interested in two parameter families then we need not consider degenerate systems forming a set of codimension 3 and so on.

So we have a *heirachy of degeneracies* by codimension and a *strategy* for their study. First we study the generic case, then degeneracies of codimension one, then two, etc. *In this study of degenerate systems we must not restrict our study to the moment of degeneracy but must include a description of the metamorphosis that takes place as the parameter passes through the critical value.*

The above general considerations are due to H. Poincaré and are applicable not only to the study of equilibrium states of evolutionary systems but to a large part of mathematical analysis in general. Although they were proposed a hundred years ago the realization of Poincaré's programme has been very slow indeed partly due to the great mathematical difficulties and partly due to psychological inertia and the dominance of the axiomatic-algebraic style.

We return, however, to equilibrium states of evolutionary systems. At the present time we can regard as solved only the problem of metamorphosis of phase curve patterns at bifurcations of equilibrium states in one parameter generic families; the case of two parameters is beyond the possibilities of present knowledge.

The results of studying generic one parameter families are summarized in Figs. 13–18. Fig. 13 depicts a one parameter family of evolutionary processes with a one dimensional phase space (the parameter ε is the horizontal axis and the state x of the system the vertical axis).

For a generic one-parameter family, the equilibrium states form a smooth curve (Γ in Fig. 13; more generally, the dimension of the manifold of equilibrium states is equal to the number of parameters). In particular this means that the bifurcations depicted on the left hand side of Fig. 14 do not occur in generic families. Small perturbations transform Γ into a smooth curve of one of the types on the right hand side of Fig. 14.

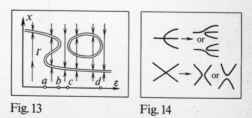

Fig. 13   Fig. 14

17

The projection of Γ onto the parameter axis for a generic one parameter family only has singularities of the fold type (for more parameters the more complex singularities of Whitney's theory appear, for instance, in generic two-parameter families the projection of the surface of equilibrium states Γ onto the parameter plane can have cusp points where three equilibrium states come together).

Thus we get singular or bifurcation values of the parameter (the critical values a, b, c, d, of the projection in Fig. 13). Away from these values the equilibrium state changes smoothly with the parameter. When the parameter value approaches a bifurcation value an equilibrium state 'dies', by combining with another one (or, going the other way, a pair is born 'out of thin air').

Of the two simultaneously appearing (or dying) equilibrium states one is stable and the other unstable.

At the moment of birth (or death) both the equilibrium states move with infinite speed: when the parameter differs from the bifurcation value by ε the distance between two nearby equilibrium states is of the order of $\sqrt{\varepsilon}$.

Fig. 15 depicts the metamorphosis of a family of phase curves on the plane for a general one parameter family. When the parameter is varied the stable equilibrium state ('node') collides with a non-stable (saddle) and then both disappear. At the moment of death a non-generic situation (a 'saddle-node') is observed.

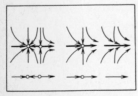

Fig. 15

In Fig. 15 it is clear that the metamorphosis is essentially one dimensional. Along the horizontal axis the same phenomena occurs as on the x axis in Fig. 13, and along the vertical axis nothing at all. Thus the metamorphosis through a saddle-node is obtained from a one dimensional metamorphosis by a 'suspension' of the ordinate axis. It turns out that all metamorphoses of equilibrium states in generic one parameter systems are such suspensions.

If stable equilibrium states describe the steady state of real systems (say in economics, ecology or chemistry) then combination with an unstable equilibrium state must cause the system to jump to a different state. When the parameter is changed the equilibrium states in the neighbourhood considered suddenly disappear. Jumps of this kind lead to the term Catastrophe Theory.

Chapter 6. Loss of Stability of Equilibrium and the Generation of Auto-Oscillations

Loss of stability of an equilibrium state on change of parameter is not necessarily associated with the bifurcation of this state. An equilibrium state can lose stability without even interacting with another state.

The corresponding metamorphosis of the phase picture on the plane is indicated in Fig. 16. Two versions are possible:

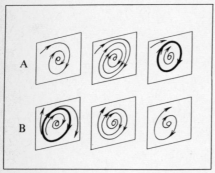

Fig. 16

A. On change of the parameter *the equilibrium state gives birth to a limit cycle* (radius of order $\sqrt{\varepsilon}$ where the parameter differs from the bifurcation value by ε). The stability of the equilibrium is transferred to the cycle, the equilibrium point itself become unstable.

B. An *unstable limit cycle collapses to an equilibrium state:* the attraction domain of the state collapses as the cycle disappears and the instability is transferred to the equilibrium state.

It was know by Poincaré and proved by Andronov and his pupils (the detailed proof was published before the war, in 1939*), that apart from the loss of stability resulting from a stable equilibrium state combining with an unstable one (as described in Chapter 5) and the A and B cases just described, for generic one parameter families of systems with two dimensional phase space no other forms of loss of stability are encountered. Later it was proved that also in systems having phase spaces of higher dimension loss of stability of equilibrium states on change of parameter must take one of the above forms (in the directions of all the additional co-ordinate axes the equilibrium states continue to be attractors).

If an equilibrium state represents a steady state in a real system the metamorphoses A and B represent the following situations.

A. On loss of stability of equilibrium the *steady state becomes a periodic oscillatory state* (Fig. 17), the amplitude of the oscillation being proportional to the square root of the criticality, the difference of the parameter from the critical value at which stability of equilibrium is lost.

This form of loss of stability is called *'soft'* loss of stability since the oscillating state for small criticality differs little from the equilibrium state.

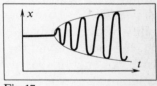

Fig. 17

* Andronov, A. A., Leontovitch, E. A.: Some cases of the dependence of limit cycles on parameters, Uchenye Zapiski Gor'kovskovo Universiteta, *6* (1939), 3–24.

    The results were also included in the first edition of the famous book of Andronov and Haikin, Oscillation Theory, Moscow 1937 (English trans: Princeton University Press, 1949).

B. Before the steady state loses stability the attraction domain becomes very small and a random perturbation can always throw the system from this domain to the other even before the attraction domain has completely disappeared.

This form of loss of stability is called 'hard'. Here the *system leaves its steady state with a jump to a different state of motion* (see Fig. 18). This state can be another stable staionary state or a stable oscillation, or some more complex motion.

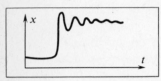

Fig. 18

Steady states of motion have recently becomes called attractors since they attract neighbouring states (transient processes). [An *attractor* is a set of attraction in phase space. Attractors that are not equilibrium states or strictly periodic oscillations are called *strange attractors* and are associated with turbulence.]

The existence of attractors with exponentially divergent phase curves on them and the stability of such situations was established at the beginning of the sixties in papers by S. Smale, D. V. Anosov, and Ja. G. Sinai on the structural stability of dynamical systems.

Independently of these theoretical works, the meteorologist E. Lorentz in 1963 described obsevations he had made in computer experiments with a fluid convection mode having an attractor in three dimensional phase space with diverging phase curves (Fig. 19) and indicated the connection between this situation and turbulence.

In Anosov's and Sinai's papers exponential divergence was established in particular for the motion of a material point on a surface of negative curvature (a saddle is an example of such a surface). The first application of the theory of

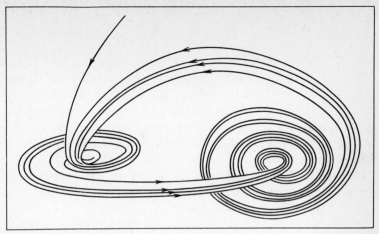

Fig. 19

exponential divergence to the study of hydrodynamic stability was published in 1966.

The motion of a fluid can be described as the motion of a material point on a curved infinite dimensional surface. The curvature of this surface is negative in many directions which leads to the rapid divergence of trajectories, i.e. to poor predictability of the flow from the initial conditions. In particular, this means that it is practically impossible to predict the longterm dynamic behaviour of weather. For a prediction 1–2 months ahead one must know the initial conditions to an accuracy of 10^{-5} times the accuracy of the forecast.

However, we return to the situation following loss of stability of an equilibrium state and assume that the state attained after the stability loss is a strange attractor (i.e. is not an equilibrium state or a limit cycle).

The transition of the system to such a state means that extremely non-periodic oscillations are observed, the precise details of which are very sensitive to small changes of the initial conditions while at the same time the average characteristics of the state are stable and do not depend on the initial conditions (for variations within some domain). An experimentor observing the motion of such a system calls it turbulent. It appears that the disorderly motion of the fluid observed on loss of stability of laminar flow with a large Reynolds number (i.e. with low viscosity) is described mathematically by such complex attractors in the phase space of the fluid. It appears that the dimension of this attractor is finite for any Reynolds number (Ju. S. Il'yashenko recently obtained an upper bound for this quantity of the order of Re^4 in the case of flows on closed 2-manifolds without boundary; for 2-manifolds with boundary the upper bound $\exp(Re^{4+\varepsilon})$ has been obtained by M. I. Vishik and A. Babin for the attractor dimension*) but tends to infinity as Re tends to infinity.

The transition from a stable equilibrium state of a process (laminar flow of a fluid) to a strange attractor (turbulence) can take place as a jump (for hard or catastrophic loss of stability) or as a soft loss of stability (Fig. 20). In the latter case the resulting stable cycle may later lose its stability. The loss of stability of a cycle in a generic one parameter family of systems can take place in a number of ways: 1) collision with an unstable cycle (Fig. 21), 2) doubling (Fig. 22) and 3) the

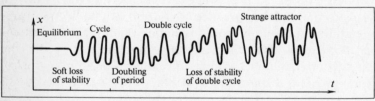

Fig. 20

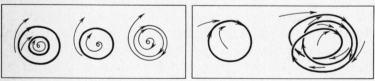

Fig. 21   Fig. 22

---

* Latest (1983) Vishik-Babin estimates are Re for 2-manifolds without boundary, $Re^{4+\varepsilon}$ for 2-manifolds with boundary.

Fig. 23   Fig. 24

Fig. 25

birth or collapse of a torus (Fig. 23) (in Andronov's terminology 'a skin comes off the cycle'). The details of these last processes depend on the resonance between the frequency of the motion along the meridians of the torus and along its axis, i.e. whether the ratio is rational or irrational. It is interesting that rational numbers with denominators 5 or higher can be regarded practically as being irrational.

The behaviour of the phase curves close to a cycle can be described approximately by an evolutionary process in which the cycle is an equilibrium state. The approximate systems arising in this way are at present studied for all cases except those close to those with strong resonance with a ratio of 1:4 (R. I. Bogdanov and E. I. Horozov). Fig. 24 shows the metamorphosis of the phase curve pattern in a plane system approximating the metamorphosis of the family of phase curves in the neighbourhood of a cycle in 3-space. It is assumed that the loss of stability occurs near a resonance of 1:3. Fig. 25 shows one of the possible sequences of metamorphoses when the ratio is close to 1:4. The basic results on this resonance are obtained not by strict mathematical arguements but by a combination of guesses and computer experiments (F. S. Berezhovskaya and A. I. Khibnik, and A. I. Neustadt).

The Poincaré-Andronov theory as described on the loss of stability of equilibrium states has so many applications in all branches of oscillation theory (to systems with a finite number of degrees of freedom and also to continuous media) that it is impossible to list them here. Systems in mechanics,

25

physics, chemistry, biology and economics lose stability all the time.

In articles on catastrophe theory soft loss of stability of equilibrium states is usually called Hopf bifurcation (partly because of me since, in talking of the Poincaré-Andronov theory to Thom in 1965, I particularly emphasised E. Hopf's work taking part of the Andronov theory over to the multidimensional situation).

In bifurcation theory as in singularity theory the fundamental results and applications were obtained without the help of catastrophe theory. The undoubted contribution of catastrophe theory was the introduction of the term attractor and the spreading of knowledge on the bifurcation of attractors. A variety of attractors are found now in all oscillatory situations. For instance it has been conjectured that different phonemes of speech are different attractors in sound-producing dynamical systems.

Even ten years ago every experimentor finding, let us say, a complex aperiodic oscillation in a chemical reaction, rejected it from consideration referring to impurity of the experiment, chance external effects and such like. Now it is clear to many that these complex oscillations can be associated with the very essence of the situation, can be determined by the fundamental equations of the problem and are not random external effects. They can and must be studied along with the classical stationary states and periodic processes of the problem.

Chapter 7. Singularities of Stability Boundaries and the Principle of the Fragility of Good Things

We consider an equilibrium state of a system depending on several parameters and assume that (in some domain of variation of the parameters) this equilibrium state does not bifurcate.

We shall describe a system by a point in the parameter space, that is by a point corresponding to the parameter value on the parameter axis (on the plane for two parameters, the parameter space for three and so on).

We consider a division of the parameter space into two parts depending on whether the equilibrium state is stable or not. Thus we obtain on the plane (in the space) of parameters the *stability domain* (consisting of those values of the parameters for which the equilibrium is stable), the *instability domain* and dividing them the *stability boundary*.

In the spirits of Poincare's general strategy (see Chap. 5) we restrict ourselves to generic families of systems depending on parameters. It turns out that the stability boundary has singularities that do not disappear with small perturbations of the family.

Fig. 26 depicts all the singularities of the stability boundary of an equilibrium state for generic two parameter families

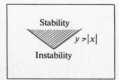

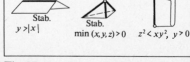

Fig. 26   Fig. 27

of evolutionary systems (with phase spaces of any dimension), Fig. 27 for three parameter families. The formulae in the diagrams describes the stability domains for a suitable choice of planar or spatial co-ordinates (which are, generally, curvilinear).

We observe that in all cases the stability domain *projects an acute angled wedge into the instability domain*. Thus for systems near the sharp part of the boundary a small perturbation is more likely to send the system into the unstable region than into the stable region. This is a manifestation of a general principle stating that good things (e.g. stability) are more fragile than bad things.

It seems that in good situations a number of requirements must hold *simultaneously* while to call a situation bad any one failure suffices. In the four parameter case we must add two more to the types described above.

As the number of parameters increase, the number of types of singularities of the stability boundaries of generic stable families rapidly grows. It was shown by L. V. Levantovski that the number of singularity types (not reducible to each other by smooth changes of parameters) remains finite for arbitrarily large numbers of parameters and the fragility of good things principle is also retained.

Chapter 8. Caustics, Wave Fronts and Their Metamorphoses

One of the most important deductions of singularity theory is the *universality* of certain simple forms like folds and cusps which are encountered everywhere and which one should learn to recognise. As well as the singularities already described one often meets some further types called the *'swallow tail'*, *'pyramid'*, *'purse'* and so on.

Suppose that a disturbance (e.g. shock wave, light or an epidemic) is being propagated in some medium.

For simplicity we start with the plane case. We suppose that at the initial moment of time the disturbance is on the curve *a* (Fig. 28) and that the speed of its propagation is 1. To find out where the disturbance will be at time *t* we must draw a segment of length *t* out from the curve along every normal. The resulting curve is called the *wave front*.

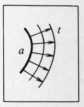

Fig. 28

Even if the initial wave front has no singularities, after some time singularities may begin to appear. For instance, for propagation inside an ellipse the singularities indicated in Fig. 29 appear. These singularities are stable (not removable by small perturbations of the initial wave front). For a gen-

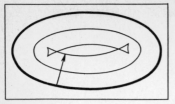
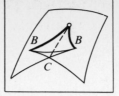

Fig. 29   Fig. 30

eric smooth initial front, as time increases, only standard singularities of this type arise.

All other singularities (e.g. the singularity at the centre of a shrinking circle) decompose on a small perturbation of the intial front into the standard ones.

For a generic three-dimensional smooth wave front only cusp ridges and the *swallow tail* standard singularities shown in Fig. 30 can appear (try to picture a wave front being propagated inside a tri-axial ellipsoid).

All the more complex singularities dissolve on small perturbations of the front into swallow tails joined by cusp ridges and by curves of self-intersection.

The swallow tail can be defined as the set of all points (a, b, c) such that the polynormal $x^4 + ax^2 + bx + c$ has a multiple root. This surface has a cusp ridge (B in Fig. 30) and a curve of self intersection (C in Fig. 30).

The swallow tail can be obtained from the spatial curve $A = t^2$, $B = t^3$, $C = t^4$. The surface is formed from all the tangents to this curve.

We consider the intersections of the swallow tail with generic parallel planes (see Fig. 31).

These intersections are plane curves. As the plane moves the curve is transformed at the moment the plane passes

Fig. 31

through the summit of the tail. The transformation (meta-morphosis) here is exactly the same as the metamorphosis of a wave front on a plane (for instance in propagation inside an ellipse).

We can describe the metamorphosis of a wave front in the plane as follows. Along with the basis space (the plane in this case) we take *space-time* (three dimensional in the present case). The propagation of a wave front in the plane is repre-sented by a surface in space-time. It turns out that this sur-face can always be regarded as a wave front in space-time (*large front*). Generically, the singularities of the large front will be swallow tails, cusp rides and self-intersections situ-ated in space-time in a general way relative to the isochrones (points at the same instant of time). Now it is easy to see which metamorphoses can take place: those of the cross-sec-tions of a large front formed by isochrones.

The study of the metamorphose of a wave front during its propagation in three-dimensional space leads similarily to the study of sections of large (three-dimensional) wave fronts in four-dimensional space time by isochrones. The resulting metamorphoses are illustrated in Fig. 32.

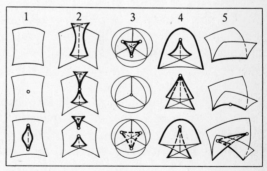

Fig. 32

The study of the metamorphoses of wave fronts was one of the problems out of which catastrophe theory arose; how-ever, even in the case of three dimensional space, catastrophe theorists have not had a great deal of success with them; Fig. 32 appeared only in 1974 when new methods (based on the crystallographic symmetry groups) were developed in singularity theory.

As well as wave fronts *systems of rays* can be used to describe the propagation of disturbances. For example the propagation of a disturbance inside an ellipse can be described using the family of internal normals to the ellipse (Fig. 33). This family has an envelope. The envelope is called a *caustic* (i.e. 'bright' since light is concentrated at it). A caustic is clearly visible on the internal surface of a cup when the sun shines on it. A rainbow in the sky is due to the caustic of the system of rays that have been completey reflected by a drop of water (Fig. 34).

Fig. 33

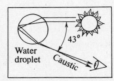

Fig. 34

The caustic of an elliptical front has four cusps. These singularities are *stable*. A nearby front has a caustic with the same singularities. All the more complex singularities of caustics resolve into the standard ones on small perturbations – cusps (given locally by $x^2 = y^3$) and self-intersections.

A system of normals to a surface in three dimensional space always has a caustic. This caustic can be constructed by marking on each normal the radius of curvature (a surface, in general, has two different radii of curvature at each point so that the normal has two caustic points).

It is not easy to describe the caustics of even the simplest surfaces; consider a tri-axial ellipsoid for instance.

Generically, caustics in three dimensional space have only the standard singularities. These singularities are the *swallow tail, pyramid* and *purse* shown in Fig. 35.

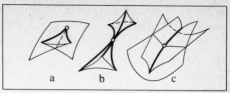

Fig. 35

The pyramid has three cusp ridges meeting at the summit. The purse has one cusp ridge and consits of two symmetrical boat bows intersecting in two lines. These singularities are stable.

All the more complex singularities of caustics in three-dimensional space resolve into these standard elements on small perturbations.

We consider an initial front (for instance a planar ellipse), its caustic and the front being propogated. It is not difficult to see that *the singularities of the propogating front slide along the caustic and fill it out.*

For instance the metamorphosis of wave front 5 in Fig. 32 corresponds to a swallow tail on the caustic. The cusp ridge moving in the threedimensional space of the wave front sweeps out the surface of the caustic (swallow tail). However, this partition of caustics into curves is not the partition of the swallow tail into curves that we encountered earlier (Fig. 31). The cusp ridge of the moving front does not have self-intersections. The cusp ridges of the moving front pass twice through the points of the lines of self-intersection of the caustic. The time interval between these passings is very small (of the order of $\varepsilon^{5/2}$ where ε is the distance from the vertex of the tail).

In exactly the same way in the metamorphoses 3 and 4 in Fig. 32 the cusp ridges of the moving front sweep out the pyramid and the purse.

If the original front is variable (under control of a parameter) then the caustic itself varies also and can undergo metamorphoses. *Metamorphoses of a moving caustic on the plane can be studied by considering sections of a large caustic in*

33

space-time as was done for fronts. The metamorphoses abtained are depicted in Fig. 36 (these metamorphoses are those of plane sections of the swallow tail, purse and pyramid). All the more complex metamorphoses decompose into sequences of these ones on small perturbation of the one parameter family.

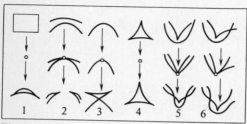

Fig. 36

We turn our attention to the metamorphosis 1 representing the formation of a caustic 'out of thin air'. The newly formed caustic has the form of a sickle with a semi-cubic cusps at the ends (*'lips'* in Thom's terminology). In a similar way a visible contour of a surface appears 'out of thin air' when the direction of projection changes (Fig. 37). Looking at a mound from above we do not see the contour. When the line of vision tilts first a point singularity appears which rap-

Fig. 37

idly grows (proportional to $\sqrt{t-t_0}$, where t_0 is the time at which the singularity appears) and takes the form of a lip. The metamorphosis described here can be regarded as the metamorphosis of a plane section of a surface with a cusp

34

ridge caused by movement of the plane (at the moment of metamorphosis the plane touches the cusp ridge – Fig. 38).

The metamorphosis 3 can also be seen as a visible contour metamorphosis, one should look at a two humped camel passing by (Fig. 39). At the moment of the metamorphosis the profile has the singularity of the curve $y^3 = x^4$.

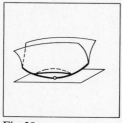

Fig. 38

Fig. 39

All the transformations of the visible contours of surfaces in generic one-parameter families are given by the first three illustrations in Fig. 36, 1–3.

The metamorphoses of caustics moving in three dimensional space are obtained as sections of large (three-dimensional) caustics in four-dimensional space-time by three dimensional isochrones. These metamorphoses are shown in Figs. 40 and 41.

One of these metamorphoses (1) describes the formation of a new caustic 'out of thin air'. We observe that that the newly formed caustic has the form of a saucer with a sharp rim. At time t after formation of the saucer the length and breadth are of the order of $\sqrt{t}$, the depth of order t and the thickness of order $\sqrt[4]{t}$.

This caustic can be seen when rays of light pass through a dispersive medium (dust, fog). V. M. Zakalyukin has suggested that observers have described these kind of caustics as flying saucers.

Cusp ridges moving in three dimensional space form the surface of a *'bicaustic'*. The singularities of a generic bicaustic

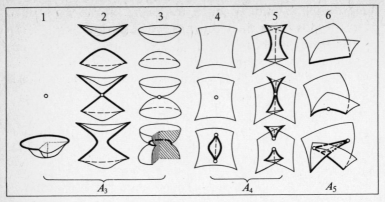

Fig. 40

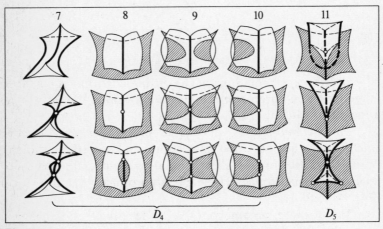

Fig. 41

corresponding to the various metamorphoses in Figs. 40 and 41 are shown in Fig. 42.

As is well known, rays describe the propogation of waves (light say) only as a first approximation. For a more precise description of a wave one must introduce a new essential parameter, the wave-length (a ray description is satisfactorily only when the wave-length is small compared with the geometric dimensions of the system).

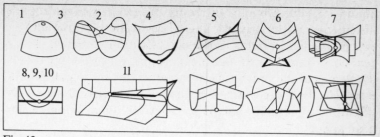

Fig. 42

The intensity of light near a caustic is great and near singularities is even greater. The coefficient of intensity amplification is proportional to $\lambda^{-\alpha}$, where λ is the wave-length and α is a rational number depending on the type of the singularity. For simple singularities the values of α are as follows:

caustic	cusp ridge	swallow tail	pyramid	purse
1/6	1/4	3/10	1/3	1/3

Thus the brightest points are at pyramid and purse singularities. In the case of a moving caustic, the singularities A_5 and D_5* (see Figs. 40 and 41) can appear producing still brighter spots momentarily ($\alpha = 1/3$ for A_5 and 3/8 for D_5).

If the light is sufficiently intensive then it can destroy the medium, starting at the point of greatest intensity and so the index α determines the relationship between the intensity of destruction and the frequency of the light.

Similar descriptions of singularities of caustics and wave fronts have been given for higher dimensional space up to dimension 10 (V. M. Zakalyukin).

This theory of singularities of caustics, wave fronts and their metamorphoses has been completely confirmed by experiments and it seems strange now that this theory was not

---

* All the singularities listed above belong to the two families A_k and D_k; see Ch. 15 for more details on these families.

obtained 200 years ago. In reality the mathematical apparatus needed is not trivial* and is associated with such diverse mathematics fields as the classification of simple Lie algebras, Coxeter's crystallgraphic groups, the theory of braids, the theory of the branching of integrals depending on parameters, etc. It is even linked (rather mysteriously) to the classification of regular polyhedra in three-dimensional Euclidean space.

Catastrophe theorists have tried to avoid the serious mathematics. For instance Zeeman's 1980 bibliography of articles on catastrophe theory omits most of the mathematical papers published from 1976 onwards. This results in catastrophe theorists trying to find answers to problems already solved by mathematicians. For instance in a 1980 paper on wind fields and the movement of ice one finds semi-successful attempts to guess the list of the metamorphoses of caustics in three-dimensional space (see Figs. 40 and 41) published by mathematicians back in 1976.

---

* The original proof of Whitney's theorem, with which we began, was about forty pages long; while one can easily understand and use the final geometrical results of singularity theory, the proofs still are long and difficult.

Chapter 9. Large Scale Distribution of Matter in the Universe

At the present time the distribution of matter in the universe is highly non-uniform (there are planets, the sun, stars, galaxies, clusters of galaxies and so on). Astrophysists nowadays suggest that in earlier stages of the development of the universe there was no such non-uniformity. How did it come about? Zel'dovich in 1970 proposed an explanation of the formation of particle clusters that is mathematically equivilent to the analysis the formation of singularities of caustics.

We consider a non-interacting particle medium, i.e. a medium sufficiently rarified that its particles pass through one another without colliding. For simplicity we assume that there are no attractions between particles and that particles move under inertia. In time t a particle starting at point x goes to the point $x + vt$.

We assume that at the initial moment of time the velocity of a particle at the point x is $v_0(x)$; the vector field v_0 is called the *initial velocity field of the medium*. As time goes by the particles move and the velocity field changes (though the velocity of each particle does not change, the vector is translated).

In Fig. 43 we show an initial velocity field v_0 of a homogeneous medium and the fields v_1, v_2, v_3 obtained from it at times $t = 1, 2, 3$. We see that the *faster particles begin to leave the slower ones behind,* with the result that the velocity field becomes three-valued at some point: here the three particles pass through the same point with different velocities.

The movement of our medium can be described as a *one parameter family of mappings of the line onto the line*. That is

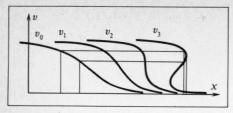
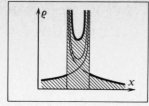

Fig. 43   Fig. 44

for each t there is a mapping g_t taking the initial position of the particle (x) to the final position $g_t(x) = x + v_0(x)t$.

The mapping g_0 is the identity mapping (the transformation that takes each point to itself). For t close to 0 the mappings are one-to-one and have no singularities. After the moment of the first overtaking, g_t has two folds.

Suppose that at the beginning the density of the medium at x was $\rho_0(x)$. The density changes with time. It is not difficult to see that after an overtaking the density graph will have the form of Fig. 44 (at a distance ε from the fold point, the density is of the order $1/\sqrt{\varepsilon}$). Thus small differences in the initial velocity field lead after long enough times to the formation of accumulations of particles (at the points of infinite density).

This conclusion still holds when one goes from a one-dimensional medium to media of any dimension and when one allows for the effects on the motion of particles due to an external or internal force field and also when relativistic effects and the expansion of the universe are allowed for.

If the force field has a potential (i.e. work done in moving along any path depends only on the initial and final points), likewise the intial velocity field, then the problem of determining the singularities of the mappings g_t and their metamorphoses with time t is mathematically identical to that of the classification of singularities of caustics and of their metamorphoses (both are within the theory of *Lagrange singularities*).

In the case of a two-dimensional medium, points of infinite density form a curve on the plane. This curve is formed by the critical values of the mapping g_t i.e. the values at the critical points (for the mapping in Fig. 1 the critical points are the points on the equator of the sphere, the critical values are the points of the visible contour on the horizontal plane).

The curve of critical values of g_t is its *caustic*. It can be defined as the set of points where two infinitely nearby rays (particle trajectories) intersect, i.e. as the usual optical caustic.

Thus the description of the metamorphoses of optical caustics carries over to the metamorphoses of particle clusters (points of infinite density of the medium) in potential motion.

The first singularity that appears in the plane is the sickle with semicubic cusps at its ends (in three dimensional space the first caustic that appears has the form of a saucer). Ja. B. Zel'dovich called such a caustic a *pancake* ('bliny' in Russian; at first pancakes were interpreted as galaxies, later as clusters of galaxies).

On subsequent motion of the medium new pancakes arise, also the existing pancakes begin to transform and may react with one another. A typical sequence of events in a two dimensional medium is illustrated in Fig. 45.

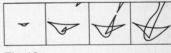

Fig. 45

All the elementary metamorphoses that are possible in a three-dimensional medium are illustrated in Figs. 40 and 41 (these results require the complicated mathematical theory of Lagrange singularities).

As a result of the metamorphoses the density has singularities of different orders on the surface of the pancakes along curves and at isolated points. We shall characterize a singularity by the *mean density* in an ε-neighbourhood of the point (i.e. the ratio of the mass in an ε-neighbourhood to the volume of the neighbourhood.)

At points on the caustic the density tends to infinity as the radius ε of the neighbourhood tends to zero.

The order of the mean density at various points of the caustic are as follows:

caustic	cusp ridge	swallow tail	purse, pyramid
$-1/2$	$-2/3$	$-3/4$	-1

With time change the singularities A_5 (mean density of order $\varepsilon^{-4/5}$) and $D_5(\varepsilon^{-1}\ln 1/\varepsilon)$ appear momentarily.

According to astrophysisists, when the radius of the universe was a thousand time smaller than at present the large scale distribution of matter in the universe was practically uniform and the velocity field was practically potential. The subsequent movement of particles led to the formation of caustics, i.e. singularities of density and clusters of particles. Up to the formation of pancakes the density remained small so that particle medium could be assumed to be non-interactive. After this the medium can be assumed to be non-interactive if neutrons account for a significant part of the mass of the universe; if, however, most of the mass is in protons and neutrons then deductions from the geometry of caustics and their metamorphoses must be treated with caution since the medium then ceases to be non-interactive.

The deductions on the clusters of matter on surfaces with the primary clusters along certain lines (cords) connecting special points (nodes) appears to agree with astronomical observations at least in general terms (S. F. Shandarin).

Chapter 10. Singularities in Optimization Problems, the Maxima Function

Many singularities, bifurcations, and catastrophes (jumps) arise in all maxima and minima problems; such problem arise, for instance, in optimization, control theory and decision theory. For instance, suppose we have to find x such that the value of a function $f(x)$ is maximal (Fig. 46). On smooth change of the function the optimum solution changes with a jump, transferring from one competing maxima (A) to the other (B).

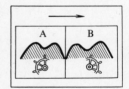

Fig. 46

Below we consider a number of problems of this type, all far from being completely solved though in some cases adequate classifications of singularities are known.

We consider a *family* $f(x, y)$ of functions of a variable x, parameterized by y. For each fixed value of y we determine the maximum of the function, denoting it by

$$F(y) = \max_{x} f(x, y).$$

The function F is continuous but not necessarily smooth even when f is a polynominal.

Example 1. Let y be the azimuth of the line of vision, x be the distance and f be the angle of elevation of the landscape at distance x in the azimuth direction y (Fig. 47). Then F determines the *line of the horizon*. It is clear that *the horizon of a smooth surface can have breaks and that these cannot be removed by small perturbations.*

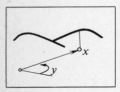

Fig. 47

The variable x and parameter y can be points in spaces of any dimension; as well as maxima functions we have minima functions.

Example 2. Let x be a point on a plane curve γ, y be a point in the region bounded by this curve and $f(x, y)$ be the distance from y to x.

We consider f as a function of the points of the curve depending on a point of the region as a parameter. Then the minimum $F(y)$ is the shortest distance from y to γ (Fig. 48). It is clear that this function is continuous but not everywhere smooth.

Fig. 48

We can consider a *spade* bounded by a curve γ and the largest possible heap of dry sand on the spade. The surface of the heap will be the graph of F. It is clear that for a generic spade the surface of the heap has a ridge (line of breaks).

The level lines of F are none other than the fronts of a disturbance propogated inside γ.

Singularity theory enables us to describe the singularities of the maxima function F in this example as well as for generic families of functions of any number of variables as long as the number of parameters y is not greater than 10 (L. N. Bryzgalova). We consider the simplest cases of one and two parameters.

By choosing suitable co-ordinates for the parameter(s) y on the axis (in the plane) and subtracting from F a smooth function of the parameter(s), we can reduce the maxima function of a generic family in a neighbourhood of any point to one of the following normal forms:

one parameter: $F(y) = |y|$;

two parameters: $F(y) = \begin{cases} |y| \text{ or} \\ \max(y_1, y_2, y_1 + y_2) \text{ or} \\ \max_x(-x^4 + y_1 x^2 + y_2 x) \end{cases}$

The formula for the one parameter case shows that, in particular, the horizon of a generic smooth landscape has no singularities other than simple breaks. The singularities of the maxima function described by the formulae for the two parameter case give (for instance, for a heap of sand on a spade) the singularities: a ridge, a point of conjunction of three ridges and the end of a ridge (see Fig. 48).

In the last case the graph of the minimum function is the part of the surface of a swallow tail (see Fig. 30) obtained by removing the pyramid (BCB) adjoining the cusp ridge (and by reflecting the surface in Fig. 30 in the horizontal plane).

For 3, 4, 5, and 6 parameters the number of different singularities is 5, 8, 12, and 17 respectively. From 7 upwards the number of non-equivalent singularity types becomes infinite, the normal forms inevitably contain 'modules' which are functions of the parameters.

All maxima (minima) functions, constructed from generic families of smooth functions are locally topologically equivalent to generic smooth functions on the parameter space (V. I. Matov).

Chapter 11. Singularities of Accessibility Boundaries

A *control system* in phase space is defined as follows: at every point of the space we have not one velocity vector (as in the usual evolutionary system) but a whole set of vectors called the *indicatrix of permissible velocities* (Fig. 49).

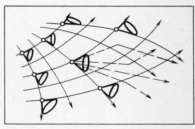

Fig. 49

The control problem is to choose at each moment of time a velocity vector from the indicatrix and by doing this to reach a target (for example to reach some subset of phase space in the shortest possible time).

The dependance of the shortest time to reach the target upon the initial point can have singularities. The singularities of the function of the minimum of the distance to a curve considered in Chap. 10 is a particular case (there the indicatrix is a circle and the target is the curve).

Unlike this particular case, singularities of the shortest times in general control problems have been studied very little. In the general case it may not be possible to reach the target under certain initial conditions. The points of the phase space from which the target can be reached (in any time) is called the accessibility domain.

The boundary of the accessibility domain can have singularities even when neither the target nor the indicatrix field have singularities. We give below a classification of the singularities of accessibility boundaries in generic controlled systems on phase planes when the indicatrix and the target are smooth curves (due to A. A. Davidov).

Of the four types of boundary singularities, three are given by the following simple formulae for appropriate choices of local co-ordinates on the plane:

1) $y = |x|$
2) $y = x|x|$
3) $y = x^2|x|$

The normal form of the fourth type of singularity contains an arbitrary smooth function B in two variables x, y and three parameters a, b and $\alpha > 1$.

4) $\begin{cases} a(x - \sqrt{y})^\alpha = x/\alpha + yB - \sqrt{y}, x \geqslant 0 ; \\ b(|x| - \sqrt{y})^\alpha = x/\alpha + yB + \sqrt{y}, x < 0. \end{cases}$

Examples of control systems and targets that have accessibility boundary singularities of these types are illustrated in Figs. 50, 51 and 52. In these diagrams the target γ is denoted by a double line, the accessibility boundary by a T-line (the spike of the T points into the accessibility domain). The curves with the arrows indicate the edges of the cones of permissible directions at each point: the horizontally hatched areas are domains of 'complete controllability' (the convex hull of the indicatrices contains 0). By looking at Figs. 50 and 52 the reader can convince himself that the singularities 1–4 are not removable.

In order to clarify this, we introduce a network of *limit curves* as follows:

At each point outside the complete controllability domain the directions of possible velocities lie inside an angle of less than 180°.

The sides of this angle determine the *limit velocities* at the point. Thus at each point outside the complete controllability domain there are two limit directions. The integral curves of the fields of limit directions (i.e. the curves that have a limit direction at every point) are called *limit curves*.

The network of limit curves is indicated in Fig. 49 together with the indicatrices of permissible velocities (being ellipses) and the angles formed by the possible directions of motion.

The boundary of the of the accessability domain consists of segments of limit curves (and possibly segments of the target when the target does not lie within the region of complete controllability, see Fig. 50). These segments meet at points which are singularities of the accessibility domain boundary.

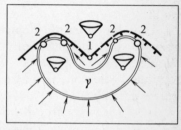

Fig. 50

In Fig. 50 the target has the form of the boundary of a smooth 'C' lying on its back. The permissible velocities at all points are the same and their directions lie within 45° with the vertical.

The inclination of all limit curves is ±45°. The accessibility boundary is denoted by *T*-like marks. Singular points on the boundary are of two types, 1 and 2.

At point 1 *two different limit curves meet*. They intersect at a non-zero angle. It is clear that for points lying above the boundary in Fig. 50 the target cannot be reached by motion in directions making angles of 45° or less with the vertical whereas for points underneath the target can be reached.

It is interesting to observe that at the vertex 1 the inaccessibility domain intrudes into the accessibility domain, the inaccessibility domain has a wedge at this point. Thus in the terms of Ch. 7 the inaccessible is good, not the accessible.

At points 2 on the accessibility boundary a segment of the limit curve meets a segment of the target. At this point the directions of the target and the limit curves coincide, hence the accessibility boundary has a tangent. The curvature of the boundary changes at point 2 with a jump from that of the limit curve to that of the target.

Now we change the target in Fig. 50 to a nearby one (nearness means nearness of tangents, curvature etc.) and change the field of the indicatrices to a nearby one. It is clear that the accessibility boundary will have a break point near 1 (where two limit curves meet at a non-zero angle). Near point 2 there will be a point with similar characteristics to those at point 2.

Thus the situation depicted in Fig. 50 is stable with respect to small perturbations. The situations depicted in Figs. 51 and 52 have a similar stability property. Events which lead to singularities of the networks of limit curves in these diagrams are as follows:

In Fig. 51 the curve K bounds the hatched complete controlability domain: in the hatched area motion in any direction is possible (if so-called mixed strategies are allowed, i.e.

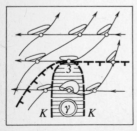

Fig. 51

motion constantly changes tack). The target in Fig. 51 lies within the domain of complete controllability. Consequently all the domain bounded by K is accessible. On the boundary K the permissible directions subtend an angle of 180°. The boundary K is formed by those points in the plane for which the double tangent convexifying the indicatrix of permissible velocities, passes through the origin of the velocity plane (a double tangent is a straight line touching the curve at two points).

In Fig. 51 the double tangent is horizontal at every point of K. At the singularity 3, *K is tangent to the double tangent to the indicatrix passing through 0.*

For a generic system such an event can only occur at isolated points of the boundary K of the complete controlability domain. In Fig. 51 this happens at point 3 where the tangent to K is horizontal.

From what has been said it is clear that the situation described is stable; small perturbations of the system, i.e. of the target and of the indicatrix field will move point 3 but not remove it.

Now we consider the set of limit curves near the point 3. Both fields of limit directions near it are smooth. We choose a system of co-ordinates such that the curves of one of the fields are straight lines. In Fig. 51 such a system has already been chosen; the first of the two families of limit curves consists of horizontal lines (directed to the left).

The curves of the second family are smooth. Along K they are tangental to the first family. At the point concerning us, 3, both families are tangent to K. From these considerations it is not difficult to see now that the net of limit curves near 3 appear as indicated in Fig. 51; above K the curves of the second family go upwards and below they go downwards (the choice of directions permits veriations which the reader can easily construct for himself).

Now in Fig. 51 it is clear that to the left of 3 the accessibility boundary follows a curve of the second field of limit directions and to the right of 3 it follows a (horizontal) curve of the first field. At 3 the two curves have a common second derivative (being a straight line and a cubic parobola). In a neighbourhood of this point the accessibility boundary is diffeomorphic to the curve $y = x^2 |x|$.*

Thus points 1 and 2 of Fig. 50 and 3 of Fig. 51 are stable examples of singularities of the first three types. The fourth type occurs in the situation shown in Fig. 52.

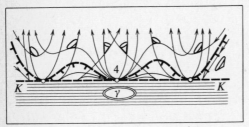

Fig. 52

Here, as in Fig. 51, the target is within the hatched domain of complete controllability. The points of the plane at which the indicatrix passes through zero lie on the boundary K of this region. It is clear that in generic control systems this event (the indicatrix passing through zero) occurs on a curve. On one side of this curve K lies the domain of complete controllability (the indicatrices include zero) and on the other side there are two limit directions. On the separating boundary these two fields coalesce (to form the field of tangents to the indicatrices at zero).

At a generic point of K the direction of this field makes a non-zero angle with K. The fourth type of accessibility boundary singularity occurs where *a tangent of K coincides with the limit direction*. In a generic system this can only occur at isolated points of K. In Fig. 52 there are three such points, one of them being the point 4. In order to study the net of limit curves in the neighbourhoods of these singular

---

* A diffeomorphism is a smooth change of variables with an inverse map that is also smooth.

points it is useful to consider our two-valued field of limit directions as a single-valued field on a surface of a two-fold covering of the plane ramified over K (using the terminology of complex analysis).

To this end we consider the set of all directed linear elements on the plane. This set is a three dimensional manifold since the direction is determined by the point at which it is attached (two co-ordinates) and its azimuth (one angular co-ordinate).

The set of all limit directions is a subset of the set of all directions. It is a smooth surface in the three-dimensional manifold of all directions. The three-dimensional manifold projects down onto the original place (each element projects onto the point to which it is attached). The surface of limit directions projects onto the part of the plane lying above K. This projection of the surface onto the plane has a singularity over K, in fact a Whitney fold.

The two-valued field of limit directions determines a single-valued directions field everywhere on the constructed surface except at the singular points of K (where the indicatrix at 0 is tangential to K) which we shall study.

The limit curves of both fields of limit directions on the constructed surface form a system of phase curves of a smooth vector field with singularities at the points we are interested in. These singular points can be nodes, foci, or saddles (in Fig. 52 the middle point is a node and the other two saddles). Thus the limit curves on the original plane are obtained from phase curves of a vector field in the neighbourhood of a singular point by a mapping which has a Whitney fold singularity.

Although the Whitney mapping and the phase curves are not completely independant (in particular, over K the phase curves are tangential to the kernel of the projection) this construction is adequate for the study of limit curves near a singular point (incidentally asymptotic curves near a parabolic curve on a surface have similar singularities).

Fig. 52 illustrates one of the variants of this situation. Here it is clear that the T-marked accessibility boundary is formed by the projections of separaticies of saddles (the outer points) under the mapping of a two-fold covering surface onto the plane. Over point 4 there is a node, into this node comes two saddle separaticies from different directions.

At the node these two curves have a common tangent and (generically) can be represented in a suitable co-ordinate system by the equations of parabolas of degree $\alpha > 1$:

$$y = A |x|^{\alpha} \ for \ x \leqslant 0, \qquad y = B |x|^{\alpha} \ for \ x > 0.$$

The fourth singularity of the accessibility boundary is obtained from this pair of α degree parabolas on the covering surface by a Whitney fold mapping. Although the curves on the covering surface and the fold mapping can be given by simple formulae there is no simple formula for the singularity of the curves obtained by the projection onto the plane.

The problem is that the system of co-ordinates normalizing the curves and the fold mapping are different. Davidov showed that the normal form of the singularity of the projected curves must contain an arbitrary function of two variables (whereas the previous three singularities have simple normal forms for suitable choices of co-ordinates).

Incidentally this situation shows the error of many catastrope theorists' common vulgarisation of R. Thom's satement that "in nature one meets only stable phenomena and therefore in every problem one should study the stable cases, rejecting the others as being unrealisable". In the present case the first three singularities are stable (with respect to diffeomorphism) whereas the fourth is not. Nevertheless all four types are encountered equally often and the study of the last no less important than of the others.

On singularities of accessibility domains, time function and optimal strategies in generic control systems with phase space of large dimension little is known, however A. A. Davidov has proved that the accessibility domain is a topological manifold with a Lipshitzian boundary.

One of the intermediate problems in the study of controlled systems is that of singularities of the convex hulls of smooth manifolds (curves, surfaces, and so on).

The *convex hull* of a set is the intersection of all half-spaces containing it. The indicatrix in a controlled system need not be convex.

However, it turns out that a non-convex indicatrix can be replaced by its convex hull. For instance the indicatrix of the velocities of a yacht blown by the wind is not convex (Fig. 53). It is possible to sail against the wind however by *tacking*, applying *mixed stategy,* that is alternating the choice of the

Fig. 53

velocity from the indicatrix. The mean velocity resulting from the mixed strategy belongs to the set of all the arithmetic means of the vectors in the indicatrix, i.e. its convex hull.

Singularities of convex hulls of generic curves and surfaces in three dimensional space have been studied by V. D. Sedykh and V. M. Zakalyukin. For curves, the hull in a neighbourhood of each point is given, up to diffeomorphism, by one of the six formulae:

$$z \geqslant 0, \quad z \geqslant |x|, \quad z \geqslant x\,|x|,$$
$$z \geqslant \min(u^4 + xu^2 + yu), \quad z \geqslant \min^2(x, y, 0),$$
$$\{z \geqslant \min^2(x, y, 0), \quad x + y \geqslant 0\}$$

(Fig. 54). For surfaces by one of the three formulae: $z \geqslant 0$, $z \geqslant |x|$, $z > \rho^2(x, y)$ where $\rho(x, y)$ is the distance of (x, y) from the angle $y \geqslant c\,|x|$ (Fig. 55). The number $c > 0$ is a modulus (invariant); hulls with different c's cannot be transformed into one another by smooth maps.

gents of order greater than 1 (called *asymptotic tangents*). For generic surfaces tangents of third order exist along curves and of fourth order at isolated points; tangents of order greater than four do not exist.

All points of a generic surface can be placed in the following seven classes according to order of tangents (Fig. 57):

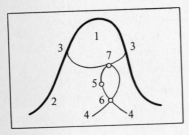

Fig. 57

1) The *domain of elliptic points* (all tangents of order 1);
2) The *domain of hyperbolic points* (two asymptotic tangents).

These two domains are divided by the common boundary:

3) The *curve of parabolic points* (one asymptotic tangent).

Inside the domain of hyperbolic points there is a special curve:

4) The *curve of inflections of asymptotic lines* (where there are third order tangents).

Finally, on this curve there are three special types of isolated points:

5) The *points of double inflection* (having fourth order tangents);

6) The *points of inflections of both asymptotic curves* (two third order tangents);

7) The *points of intersection of the curves* 3) *and* 4). For generic surfaces, at points of type 6) two branches of the

curve of inflections meet at a non-zero angle while at points of type 7) the curves 3) and 4) are tangential (of first order).

This classification of the points of a surface (due to O. A. Platonova and E. E. Landis) is connected in the following way with singularities of wave fronts.

Mathematicians can regard objects of any form as *points*. Consider for example the set of all non-vertical straight lines on the plane (x, y).

Such lines are determined by equations of the form $y = ax + b$. So a single line is determined by the pair (a, b) and can be regarded as a point of the plane with co-ordinates (a, b). This plane is called the *dual plane* of the original plane. Its points are the lines of the original plane.

A smooth curve on the original plane has a tangent at each of its points, as a point moves along the curve the tangent changes and consequently the point moves in the dual plane. Thus in the dual plane there is a curve that represents the set of all the tangents of the original curve. This curve is called the *dual curve*.

If the original curve is smooth and convex then the dual curve is smooth. If the original curve has a point of *inflection* then on the dual curve there is a corresponding *cusp* (Fig. 58).

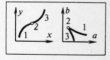

Fig. 58

Curves dual to generic smooth curves have the same singularities as generic wave fronts on the plane and the same metamorphoses result from a smooth deformation of the original curve as occur in the propagation of a generic wave front in the plane.

In exactly the same way the planes in three-dimensional spce form a *dual three-dimensional space* and all the tangent planes to a smooth surface form a *dual surface*. The singularities of a surface dual to a generic surface are the same as those of a generic wave front, i.e. cusp ridges with swallow tails.

Curves of parabolic points on the original surface correspond to cusp ridges on the dual surface. Special points on this curve where it touches curves of asymptotic inflections correspond to swallow tails. The curves of self intersection of the swallow tail consist of double-tangent planes to the original surface. Consequently at the point 7 the two tangential points coalesce and so a one parameter family of double-tangent planes terminates.

The different classes of points on a generic surface are also seen in the singularities of a visible contour. If the direction of projection is generic then, by Whitney's theorem, the only singularities are folds and cusps. By choosing the direction of projection in a special way, however, one can obtain certain non-standard projections of a generic surface. It turns out that all such projections reduce locally to the projection of one of the following nine surfaces $z = f(x, y)$ along the x-axis:

type	1	2	3,4	5
f	x^2	$x^3 + xy$	$x^3 \pm xy^2$	$x^3 + xy^3$

6	7	8,9
$x^4 + xy$	$x^4 + x^2 y + xy^2$	$x^5 + x^3 y \pm xy$

(the surface are projected onto the plane (y, z), the reduction being realized by a change of co-ordinates $X(x, y, z)$, $Y(y, z)$, $Z(y, z)$).

The visible contours corresponding to these projections are illustrated in Fig. 59.

The correspondence between the classification of projections and the points on a surface is as follows. Type 1 is a projection along a non-asymptotic direction (Whitney fold).

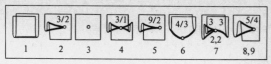

Fig. 59

Projections along asymptotic directions at generic points of the hyperbolic domain are of type 2. This projection has a Whitney cusp singularity. Under a small perturbation of the projecting direction the singular point moves only slightly on the surface; the new direction turns out to be asymptotic at a nearby point. *So to see a cusp one has only to look at any surface along an asymptotic direction.* On movement of the surface or of the observer the singularities 3, 4, and 6 appear momentarily.

The projections 6 (and 8 or 9) correspond to the hyperbolic domain (and asymptotic tangents of third and fourth orders respectively).

On the spine of a two humped camel (see Fig. 39) there is a curve of inflection of asymptotics. The third order tangents passing through its points form a surface. As the camel passes by we intersect this surface twice. At the moment of intersection the visible contour of the spine has a singularity of the type $y^3 = x^4$ and the projection is of type 6.

The remaining singularities arise from projections along directions that are asymptotic at parabolic points. The simplest of these are the singularities 3 and 4. The projection 3 occurs at the moment when a contour is first observed on descending a mound (see Fig. 37). The first contour point seen is parabolic.

Singularity 4 occurs when two components of a visible contour come together or split up.

Singularities 5, 7, 8 and 9 only arise from isolated projecting directions and one needs to deliberately search for them. (8 and 9 come from projections along tangents of fourth order, 7 along a parabolic tangent of third order and 5 a point of "parallelism of asymptotic directions at infinitely close parabolic points"). For projections from isolated points there are four more singularities: $z = x^3 \pm xy^4$, $z = x^5 + xy$ and $z = x^4 + x^2 y + xy^3$. Thus the total number of nonequivalent singularities of generic smooth surface projections from all points of the ambient three space is 14 (O. A. Platonova and O. P. Sherbak).

Chapter 13. Problems of By-Passing Obstacles

We consider an obstacle in three-dimensional space bounded by a smooth surface (Fig. 60). It is clear that the shortest path from x to y avoiding the obstacle consists of line segments and segments of geodesics (curves of minimal length) on the obstacle surface. The geometry of the shortest paths is greatly affected by the inflections of the obstacle surface.

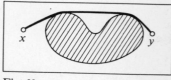

Fig. 60

We assume that the starting point x of the paths is fixed and consider the shortest paths to all possible points y. These paths start with straight line segments that are tangent to the obstacle. The following segments form a bundle (one parameter family) of geodesics on the obstacle surface. The next segments are more line segments tangent to these geodesics. These can then either terminate at the end point y or again be tangent to the surface, and so on.

We consider the simplest case of paths consisting of initial and final tangent segments and a single geodesic segment in between. Neighbouring geodesic segments in the bundle form a domain on the obstacle surface. At each point of this domain there is determined a direction of the bundle geodesic. At generic points this direction is not asymptotic. The condition that a bundle geodesic should have an asymptotic tangential direction is one condition for points on the surface. For a generic surface and bundle this condition is satisfied along some surve (depending on the bundle) lying on the surface. In Fig. 61 the asymptotic directions are indicated by horizontal line segments, the curve of tangencies by the letter K and the geodesics by dark curves.

At isolated points (such as 0 in Fig. 61) the curve K itself has an asymptotic direction. These points are the intersection of K with the curve 4 of the inflections of the asymptotes (see Chap. 12).

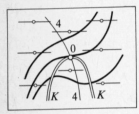

Fig. 61

Thus we get a *two parameter family of paths:* one parameter determining the geodesic in the bundle and the other the point at which the tangent segment leaves the surface. Along each path a time function is defined (calculated from the starting point x). The time taken to reach the end point y is not uniquely determined (several paths can lead to y) and not all paths by-pass the obstacle. Nevertheless it is clear that study of the multivalued time function defined above is a necessary stage in the study of singularities of systems of shortest paths.

We put another generic surface (a wall) behind the obstacle and consider the *break mapping* of the surface of the obstacle onto the wall, mapping each point of the obstacle to the point of intersection with the wall of the tangent breaking away from the bundle geodesic at that point.

As the wall goes off to infinity the break mapping become the *Gaussian mapping of the bundle;* each point on the obstacle surface corresponds to a point on the unit sphere, namely the end of the vector of unit length parallel to the tangent to the geodesic.

The break and Gaussian mappings of bundles have singularities along the curve where the geodesic directions are asymptotic. These singularties turn out to be folds at general points and cusps at special points where the direction of the curve is itself asymptotic (O. A. Platonova).

The multivalued time function also has singularities at points corresponding to an asymptotic break. For a suitable choice of smooth coordinates the time function has the form $T = x - y^{5/2}$ in the neighbourhood of a general singular point of the surface $y = 0$. In other words if we take points on the breaking-away rays at a distance T from the initial point, then those points form the surface of a front with a cusp ridge given locally by $x^2 = y^5$ (Fig. 62).

An analogous result is obtained for the plane case – the front is now called the evolvent and has a singularity of the form $x^2 = y^5$ at points of the inflection tangent (Fig. 63).

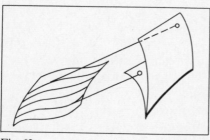

Fig. 62

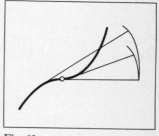

Fig. 63

Only qualitative results are known on the structure of singularities of the time function and of the fronts in the three-space obstacle problems at the singular points (at cusp points of the Gaussian map of the bundle). Even less is known on the shortest path length function when more than two switchings occur (more than one geodecsi segment is involved).

Here, as in other problems of singularity theory, much remains to be done.

Chapter 14. Symplectic and Contact Geometries

Many problems in singularity theory (for instance, the classifications of singularities of caustics and of wavefronts and also of determining the singularities in optimization and variational calculus problems) are understandable only in terms of the geometries of symplectic and contact manifolds. These geometries are pleasantly different from the three usual ones (those of Euclid, Lobachevskii and Riemann).

We begin with three examples of singularities of a special type.

1. The Gradient Mapping

Suppose that we have a smooth function on Euclidean space. Then the *gradient mapping* is the mapping that associates to each point the gradient of the function at the point. Gradient mappings are a very special class of mappings between spaces of the same dimension.

The study of generic gradient mapping singularities is not the same as the study of singularities of generic mappings between spaces of the same dimension; the gradient singularities number is "less" in the sense that not all singularities classes admit gradient representatives, but also "more" because phenomena that are not typical for general mappings can be typical for gradient mappings.

2. The Normal Mappings

We consider the set of all vectors normal to a surface in three dimensional Euclidean space. We associate to each vector its end point (the vector p based at the point q goes to the point

$p+q$). We obtain a mapping of the three dimensional manifold of normal vectors into a three dimensional space (from n dimensions to n dimensions, if we start with a submanifold of any dimension in n-dimensional Euclidean space).

This mapping is called the *normal mapping* of the original manifold. The singularities of normal mappings of generic submanifolds form a special class of singularities of mappings between spaces of the same dimension. The critical values of a normal mapping form a caustic (the locus of the original submanifold centres of curvature, c.f. Fig. 33 where original submanifold is an ellipse).

3. *The Gaussian Mappings*

We consider a two-sided surface in three dimensional Euclidean space. We transfer the unit positive normals from all points to the origin. The ends of these vectors lie on a unit sphere. The resulting mappings of the surface onto the sphere is called the *Gaussian mapping*.

Gaussian mappings constitute yet another special class of mappings between manifolds of the same dimension (which is $n-1$ if we start with a hypersurface in an n dimensional space.)

It turns out that the *typical singularities of all these three classes* (gradient, normal and Gaussian) *are identical:* all three theories are special cases of the general theory of Lagrangian singularities in symplectic geometry.

Symplectic geometry is the geometry of phase space (the position and momentum space of classical mechanics). It appears throughout the long development of mechanics, variational calculus, etc.

In the last century this branch of geometry was called analytical dynamics and Lagrange took pride in abolishing pictures from it. To penetrate symplectic geometry it is simplest to bypass the long historical route and use the axiomatic method (as was noted by Bertrand Russell this method has many advantages, similar to the advantages of stealing over honest work).

The essence of this method lies in turning theorems into definitions, the content of a theorem then becomes the motivation for a definition, which algebraists usually omit for the sake of their art glory (to understand a definition without its motivation is impossible, but how many passengers in an aeroplane understand how and why it is made?).

Pythagoras's theorem, a supreme achievement of mathematical culture in its time, is relegated by contemporary axiometic treatment of Euclidean geometry to a hardly distinguishable definition: *Euclidean structure* in linear space is a symmetric function (scalar product) of a pair of vectors, linear in each argument, for which the product of a non-zero vector with itself is positive.

The definition of *symplectic structure* in linear space is analogous. It is a skew-symmetric function (skew-scalar product), linear in each argument and non-degenerate (no non-zero vector is skew orthogonal to every other vector; that is, its skew-scalar products with some vectors are non-zero).

Example: We will take as the skew-scalar product of two vectors on an oriented plane the signed area of the parallelogram formed by them (that is, the determinant of the matrix consisting of the components of the vectors). This product is a symplectic structure on the plane.

Three dimensional space (and generally any odd dimensional spaces) admit no symplectic structure. Symplectic structure in four dimensional space (and in general in even dimensional space) is easy to construct by considering the space as a sum of two dimensional planes. The skew-scalar product decomposes into the sum of the areas of the projections onto the planes.

All symplectic spaces of the same dimension are isomorphic (as are all Euclidean spaces). We will call the skew-scalar product of two vectors the "area" of the parallelogram defined by them.

Each linear subspace of a linear space has an *orthogonal complement* of dimension equal to the codimension of the original subspace.

In symplectic space there is a *skew-orthogonal complement* to every linear subspace, consisting of all vectors for which the skew-scalar products with all vectors of the subspace are zero. The dimension of this skew-orthogonal complement is also equal to the codimension of the original subspace.

A linear subspace which is its own skew-orthogonal complement is called a *Lagrangian subspace*. Its dimension equals one half of the dimension of the original subspace.

A *Riemannian structure* on a manifold is determined by a Euclidean structure on the tangent space to the manifold at each point.

In exactly the same way a *symplectic structure on a manifold* is given by a symplectic structure on each tangent space; however, unlike the Riemannian case these structures are not arbitrary but are concordant, as is explained below.

A Riemannian structure on a manifold allows us to measure the lengths of its curves by summing the lengths of small vectors which form the curves. In exactly the same way, symplectic structures enable us to measure the "area" of an orientable two dimensional surface lying in a symplectic manifold by taking the sum of the "areas" of the small parallelograms forming it. The additional condition connecting symplectic structures in different tangent spaces is this: the "area" of the whole boundary of any three dimensional figure is zero.

One can introduce the structure of a symplectic manifold into any linear symplectic space by defining a skew-scalar product of the vectors at each point as the skew-scaler product of the vectors translated to the origin. It is easily checked that the concordance condition is fulfilled.

There exist many non-isomorphic Riemannian structures in the neighbourhood of a part of the plane or higher dimensional space (to distinguish them Riemann introduced his curvatures).

Unlike Riemannian manifolds, *all symplectic manifolds of a given dimension are isomorphic in a neighbourhood of each point* (one can be mapped onto another by an area-preserving mapping. Thus each symplectic manifold is locally iso-morphic to a standard symplectic linear space. Into such a space can be introduced co-ordinates $(p_1, \ldots, p_n, q_1, \ldots, q_n)$ such that the skew-scalar product equals the sum of the signed areas of the projections onto the planes $(p_1, q_1), \ldots, (p_n, q_n)$.

A submanifold of a symplectic space is called a *Lagran-gian manifold* if the tangent space at each point is Lagran-gian.

A fibration of a symplectic manifold is called a *Lagran-gian fibration* if all the fibers are Lagrangian manifolds. Any Lagrangian fibration is locally isomorphic to a standard fi-bration of the phase space over the configuration space $(p, q) \rightarrow q$ (the fibres are the spaces of momenta with $q = $ con-stant). The configuration space (that is, the q-space) is called the *base space* of the fibration.

We assume now that in the space of the Lagrangian fibra-tion there is given another Lagrangian manifold. Then there is a smooth mapping of this Lagrangian manifold to the base of the Lagrangian fibration (that is, on the configuration space with co-ordinates q_i): to each point (p, q) of the La-grangian manifold is associated a point q in the configura-tion space.

The resulting mappings between manifolds of the same dimension n are called *Lagrangian mappings* and their singu-larities are called *Lagrangian singularities*. These form a spe-cial class of singularities of smooth mappings between mani-folds of the same dimensions. For this class there is a classifi-cation theory analogous to the general theory of singularities. For $n = 2$, generic Lagrangian singularities are only folds and cusps as with general singularities; however, here there are two versions of the cusp which are not Lagrangian equival-ent.* In the three-dimensional case there are singularities of

---

* A *Lagrangian equivalence* of two Lagrangian singularities is a mapping from one Lagrangian fibration space to another which maps fibres onto fibres, the first symplectic structure onto the second and the first Lagrangian submanifold onto the second.

generic Lagrangian mappings which do not occur in generic non-Lagrangian mappings.

Now we show that *gradient maps, normal maps and Gaussian maps are Lagrangian:*

1. Let F be a smooth function of p. Then the manifold $q = \partial F / \partial p$ is Lagrangian. Therefore singularities of gradient maps are Lagrangian.

2. We consider a smooth submanifold in Euclidean space. We consider the set of all vectors orthogonal to it (at all points q). The manifold consisting of vectors p attached to the points $p + q$ is Lagrangian. The normal mapping can be considered as a Lagrangian mapping from this manifold to the base $(p, p + q) \rightarrow (p + q)$.

3. We consider the manifold of all directed straight-lines in Euclidean space. This manifold is symplectic since it can be considered as the phase space of a point moving on a sphere (the direction of the line determines a point on the sphere and the point of intersection of the line with the tangent plane of the sphere perpendicular to it gives the magnitude of the momentum).

We look at the manifold of all directed normals to a surface in our space. This submanifold in the symplectic manifold of directed lines is Lagrangian. Its Gaussian mapping can be considered as a Lagrangian mapping (a projection mapping from the submanifold just constructed to the sphere which is the base of a Lagrangian fibration of the phase space).

Thus the theory of gradient, normal and Gaussian singularities reduces to the theory of Lagrangian singularities.

The symplectic structure in the manifold of directed lines that we have considered is not so artifical a formation as it seems at first sight. The point is that the solution set for any variational problem (or, more generally, the solution set for the Hamilton equation with a fixed value of the Hamiltonian) forms a symplectic manifold, which is very useful for investigating the properties of these solutions.

We consider, as an example, the two parameter family of rays leaving the geodesics of the obstacle surface in three-dimensional space, as shown in Fig. 62. This family is a two-dimensional Lagrangian subvariety in the four-dimensional space of all rays. But, unlike the Lagrangian submanifold we encountered earlier, this one is a Lagrangian variety which itself has singularities. These singularities occur at the points where the leaving rays are asymptotic to the obstacle surface. Such rays form a cusp ridge (of the type $x^2 = y^3$) in the Lagrangian manifold of all leaving rays.

On this cusp ridge there are special points, in the neighbourhood of which the manifold of the leaving rays is symplectic diffeomorphic to an open swallow tail. The open swallow tail is the surface in the four-dimensional space of all polynomials $x^5 + ax^3 + bx^2 + cx + d$ consisting of polynomials with roots of multiplicity at least three.

This space of polynomials has a natural symplectic structure inherited from the $SL(2, R)$-invariant structure of the binary forms space and the open swallow tail is a Lagrangian subvariety in the polynomials space. The open swallow tail surface is also encountered in other problems of singularity theory (for example, in the investigation of caustics sweeping by the cusp ridges of the moving wave fronts) and is probably one of the principal examples for a future theory of Lagrangian varieties with singularities.

In Euclidean and Riemannian geometry there is an extensive theory of external curvature. In addition to the intrinsic properties of the submanifold determined by its metric, there are differences in the ways that manifolds with the same intrinsic properties can be embedded.

In symplectic geometries, as recently shown by A. B. Guivental, the case is simpler: the intrinsic geometry (the restriction of the symplectic structure to the set of tangent vectors to a submanifold) determines the external properties. In other words, *submanifolds with the same intrinsic geometry can be locally transformed one in to an other by a diffeomorphism which preserves the symplectic structure of the ambient space.* This opens up a new chapter in singularity theory, the investigation of singularities of the embeddings of submanifolds in a symplectic space, the importance of which was noted by R. Melrose in recent papers on diffraction. The beginnings of a classification of such singularities is obtained, via Guivental's theorem, from the results of J. Martinet and his followers, on degenerations of the symplectic structures. For example, a generic two-dimensional submanifold in four-dimensional symplectic space can be mapped locally by a symplectic structure preserving transformation to one of two normal forms:

$$p_2 = q_2 = 0 \quad \text{or} \quad q_1 = 0, \ p_2 = p_1.$$

On odd dimensional manifolds there can be no syplectic structures, but there are instead contact structures. Contact geometry does for optics and the theory of wave propagation what sympectic structures do for mechanics.

A contact structure on an odd-dimensional manifold is determined by choosing a hyperplane (subspace of codimension 1) in the tangent space at each point. Two fields of hyperplanes on manifolds of fixed dimension are locally equivalent (one can be transformed to another by a diffeomorphism) if they are both generic near the points under consideration.

A *contact structure* is a field of hyperplanes on an odd-dimensional manifold, the fields being generic at each point of the manifold.

The manifold of all linear elements in the plane is a contact manifold; it is three-dimensional. Its contact structure is given as follows: the velocity of an element belongs to the (hyper)plane field if the velocity of the point of application belongs to the element. The same definition gives a contact structure to the $(2n-1)$-dimensional manifold of tangent hyperplanes on every n-dimensional manifold.

The role of Lagrangian manifolds in contact cases passes to Legendrian manifolds (integral submanifolds of the field of hyperplanes of greatest possible dimension, that is, of dimension m in a contact manifold of dimension $2m+1$).

The singularities of wave fronts, Legendre transforms and also of the hypersurfaces that are dual to smooth ones are all

Legendrian singularities. The whole of symplectic theory (including for instance, Guivental's theorem) has contact analogues, which are extremely useful when investigating singularities in variational problems.

In recent years symplectic and contact geometries have encroached on all areas of mathematics. As each skylark must display its comb, so every branch of mathematics must finally display symplectisation. In mathematics there exist operations of different levels: functions acting on numbers, operators acting on functions, functors acting on operators, and so on. Symplectisation belongs to the small set of highest level operations, acting not on details (functions, operators, functors), but on all the mathematics at once. Although some such highest level operations are presently known (for example, algebraisatin, Bourbakisation, complexification, superisation, symplectisation) there is as yet no axiomatic theory describing them.

Chapter 15. The Mystics of Catastrophe Theory

The applications of singularity theory to the natural sciences are not the only aspects of catastrophe theory; along with the concrete investigations of the Zeeman type one has the more philosophical work of the mathematician René Thom who first revealed the universality of the Whitney singularity theory (and the preceding work of Poincaré and Andronov on bifurcation theory), introduced the term 'catastrophe theory' and has been a great propagandist for the subject.

A particular aspect of Thom's work on catastrophe theory is his original style: he established a fashion in not giving even sketchy formulations of results, let alone proofs. Zeeman, while admiring this style, points out that the meaning of Thom's words only becomes clear after inserting 99 of your lines between every two of Thom's.

In order to demonstrate this style the following is an extract of a survey on perspectives in catastrophe theory given by Thom in 1974:

Sur le plan de la philosophie proprement dite, de la métaphysique, la théorie des catastrophes ne peut certes apporter aucune réponse aux grands problèmes qui tourmentent l'homme. Mais elle favorise une vision dialectique, héraclitéenne de l'univers, d'un monde qui est le théâtre continuel de la lutte entre 'logoi', entre archétypes. C'est à une vision fondamentalement polythéiste qu'elle nous conduit: en toutes choses, il faut savoir reconnaitre la main des Dieux. Et c'est peut-être là aussi qu'elle trouvera les limites inéluctables de son efficacité pratique. Elle connaitra peut-être le même sort que la psychanalyse. Il ne fait guere de doute que l'essentiel des découvertes de Freud en psychologie ne soit vrai. Et cependant, la connaissance même de ces faits n'a eu que très peu d'efficacité sur le plan pratique (pour la cure des troubles mentaux, notamment). De même que le héros de l'Iliade ne pouvait s'opposer à la volonté d'un Dieu, tel Poséidon, qu'en invoquant le pouvoir d'une divinité opposée, telle Athéna, de même nous ne pourrons restreindre l'action d'un archétype qu'en lui opposant un archétype antagoniste, en une lútte ambigue au résultat incertain. Les raisons mêmes qui nous permettent d'étendre nos possibilités d'action en certain cas nous condamneront à l'impuissance en d'autres. On pourra peut-être démontrer le caractère inéluctable de certaines catastrophes, comme la maladie ou la mort. La connaissance ne sera plus nécessairement une promesse de réussite, ou de survie; elle pourra être tout aussi bien la certitude de notre échec, de notre fin.

(Thom, R.: 'Catastrophe Theory: its present state and future perspectives' in: Dynamical Systems, Warwick, 1974, Lecture Notes in Math., *468,* Berlin – Heidelberg – New York, 1975, p. 372.)

The nice results of singularity theory are happily not dependant upon the dark mystics of catastrophe theory. But in singularity theory, as in all mathematics, there is an element of the mysterious: mathematical objects and theories, which at first seem quite independent, turn out to be closely related.

One example of such a relationship is the enigmatic (though partly understood) A, D, E-classification. It occurs in such diverse areas of mathematics as, for example, critical points of functions, Lie algebras, categories of linear spaces, caustics, wave fronts, regular polyhedra in three-dimensional space and the Coxeter crystallographic reflection groups.

Common to all these is the requirement of *simplicity* or the *absence of moduli.* Simplicity means the following: Each classification is a partition of some space of objects into classes. An object is called *simple* if all objects in some neighbourhood belong to a finite set of classes.

Example 1. We shall say that two sets of lines passing through the origin 0 in the plane are *equivalent* if one can be transformed into the other by a linear transformation of the type $(x, y) \rightarrow (ax + by, cx + dy)$. Any such set of three lines is simple (any set of three distinct lines is equivalent to $x = 0$, $y = 0$, $x + y = 0$). Any such set of four lines is not simple (prove it!).

Example 2. We classify the critical points of (complex-valued) smooth functions by placing functions in the same class if one can be transformed into another by a smooth (complex) change of the local coordinates. *A list of the simple singularities* (say for functions of three variables) *consists of two infinite series plus three exceptional cases:*

$$A_k = x^2 + y^2 + z^{k+1}, \; k \geqslant 1; \quad D_k = x^2 + y^2 z + z^{k-1}, \; k \geqslant 4;$$
$$E_6 = x^2 + y^3 + z^4, \quad E_7 = x^2 + y^3 + yz^3, \quad E_8 = x^2 + y^3 + z^5.$$

Example 3. A quiver is a set of points together with some arrows joining them. If to every point there is associated a linear space (point, straight line, plane) and to every arrow a linear map (from the space associated with the beginning of the arrow to the one at the end), then we say that a quiver representation is given. Two representations are *equivalent* if one can be transformed into the other by linear transformations.

In Fig. 64 the left quiver is simple, the right one is not (see Example 1).

It is known that all *the connected simple quivers are precisely those which are obtained from the Dynkin diagrams (Fig. 65) choosing arbitrary arrow directions on its segments.* The diagrams involved form two infinite series and three exceptional cases.

Simple singularites of caustics and wave fronts also form two infinite series A_k and D_k and three exceptional cases E_k (the lowest members of the series are shown in Figs. 30–41).

The symmetry groups of regular polyhedra in three-dimensional space similarly exhibit the two infinite series plus three special cases (the special cases are the symmetry groups of the regular tetrahedron (E_6), octahedron (E_7) and icosahe-

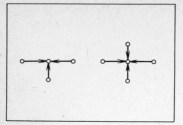

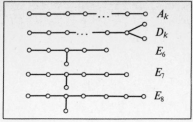

Fig. 64   Fig. 65

dron (E_8) the infinite series corresponding to symmetry groups of regular polygons and regular dihedra, i.e. two-sided polygons with faces of different colours).

Functions, quivers, caustics, wave fronts and regular polyhedra do not seem to have much in common. In fact corresponding objects do not bear the same label purely by change. For example, from the icosahedron one can get the function $x^2 + y^3 + z^5$, from this function topology one can get the Dynkin diagram E_8 and also to the corresponding caustic and wave front.

To easily checked properties of one of a pair of associated objects may correspond properties of the other which are not evident at all. Thus the relations between all the A, D, E-classifications are used for the simultaneous study of all simple objects regardless of the fact that many of them (such as the connections between functions and quivers) remain an unexplained manifestation of the mysterious unity of the universe.

Singularity theory interpretations are at present (1983) known for all Coxeter reflections groups, including the noncrystallographic ones, like H_3 and H_4. For instance, H_3, the icosahedron's symmetries group, is related to the evolvents of plane curve metamorphoses at the curve's inflection points. In the corresponding plane obstacle problem the time function's graph is diffeomorphic to the H_3 singular orbits variety (and also to the union of all tangents of the curve $x = t, \ y = t^3, \ z = t^5$). In the 3-space obstacle problem the same variety describes the front singularity at some obstacle points.

References

Arnold, V. I., Varchenko, A. N., Gasein Zade, S. M.: Singularities of differentiable mappings, "NAUKA", Moscow 1981

Arnold, V. I.: Singularities of ray systems, Russian Mathematical Surveys, *38* No. 2 (1983)

Arnold, V. I.: Singularities in the calculus of variations. In: Contemporary problems of mathematics, vol. 22, Moscow, VINITI, Itogy nauki i techniki, 1983

Breaker, T., Lander, L.: Differential forms and catastrophes, „MIR", Moscow 1977 (Translation of a Western book)

Golubitskii, M., Giiemin, V.: Stable mappings and their singularities, "MIR", Moscow 1980

Poston, Stewart: Catastrophe theory and its applications, "MIR", Moscow 1980 (Translation of a Western book)

Whitney, H.: Mappings of the plane into the plane, Ann. Math. 1955, *62,* p. 374–410

Thom, R.: Stabilité structurelle et morphogénèse, N.Y. 1972

Zeeman, E. C.: Catastrophe theory: selected papers (1972–1977), Mass. 1977

V. I. Arnold

Geometrical Methods in the Theory of Ordinary Differential Equations

Translated from the Russian by J. Szücs
English translation edited by M. Levi

1983. 153 figures. XI, 334 pages
(Grundlehren der mathematischen Wissenschaften, Band 250)
ISBN 3-540-90681-9

Contents: Special Equations. – First-Order Partial Differential Equations. – Structural Stability. – Perturbation Theory. – Normal Forms. – Local Bifurcation Theory. – Samples of Examination Problems.

Written by one of the world's most famous analysts, this book develops a series of basic ideas and methods for the investigation of ordinary differential equations and their applications in the natural sciences. One pervading feature is the use of elementary methods of integration from the viewpoint of general mathematical concepts such as resolutions of singularities, Lie groups of symmetries, and Newton diagrams. At the center of the investigations is the qualitative theory of differential equations (structural stability, Anosov systems), asymptotic methods (averaging, adiabatic invariants), the analytic methods of the local theory in the neighborhood of a singular point or a periodic solution (Poincaré normal forms), and the theory of bifurcations of phase portraits in the variation of parameters. First order partial differential equations are investigated with the help of the geometry of a contact structure.
Arnold's book is addressed to a wide group of mathematicians applying differential equations in physics, the natural siences and engineering.

Springer-Verlag
Berlin
Heidelberg
New York
Tokyo

V. I. Arnold

Mathematical Methods of Classical Mechanics

Translated from the Russian by K. Vogtmann, A. Weinstein

1978. 246 figures. X, 462 pages
(Graduate Texts in Mathematics, Volume 60)
ISBN 3-540-90314-3

Contents: Newtonian Mechanics: Experimental Facts. Investigation of the Equations of Motion. – Lagrangian Mechanics: Variational Principles. Lagrangian Mechanics on Manifolds. Oscillations. Rigid Bodies. – Hamiltonian Mechanics: Differential Forms. Symplectic Manifolds. Canonical Formalism. Introduction to Perturbation Theory. – Appendices.

V. I. Arnold

Gewöhnliche Differentialgleichungen

Übersetzt aus dem Russischen von B. Mai

1980. 259 Abbildungen. 275 Seiten
ISBN 3-540-09216-1

Inhaltsübersicht:
Grundbegriffe. – Grundlegende Sätze. – Lineare Systeme. – Beweise der grundlegenden Sätze. – Differentialgleichungen auf Mannigfaltigkeiten. – Prüfungsprogramm. – Beispiele für Prüfungsaufgaben. – Einige häufig benutzte Bezeichnungen. – Sachverzeichnis.

Mit dieser Übersetzung liegt eines der bekanntesten elementaren Lehrbücher über Differentialgleichungen erstmals auch in deutscher Sprache vor. Im Zentrum der Darstellung stehen zwei Fragenkomplexe: der Satz über die Begradigung eines Vektorfeldes, der den gewöhnlichen Sätzen von der Existenz, der Eindeutigkeit, der Differenzierbarkeit der Lösungen äquivalent ist und die Theorie der linearen autonomen Systeme. Auf die Anwendung der gewöhnlichen Differentialgleichungen in der Mechanik wird ausführlicher als allgemein üblich eingegangen. So wird bereits auf den ersten Seiten die Bewegungsgleichung des Pendels aufgestellt, an deren Beispiel die Effektivität der eingeführten Begriffe und Methoden im folgenden jedesmal nachgeprüft wird. Durch Hervorheben der geometrischen, qualitativen Seite der untersuchten Probleme gelingt es dem Autor, dem Leser zahlreiche fundamentale Begriffe, die bei der klassischen Koordinatenschreibweise im Verborgenen bleiben, nahezubringen.
An Vorkenntnisen benötigt der Leser lediglich den Stoff der Standardvorlesungen über Analysis und lineare Algebra.

Springer-Verlag
Berlin
Heidelberg
New York
Tokyo